The INSIDE GUIDE

STEM SUPERWOMEN

Ada Lovelace

By Joyce Jeffries

Cavendish Square

Published in 2026 by Cavendish Square Publishing, LLC
2544 Clinton Street Buffalo, NY 14224

Website: cavendishsq.com

Library of Congress Cataloging-in-Publication Data

Names: Jeffries, Joyce, author.
Title: Ada Lovelace / Joyce Jeffries.
Description: Buffalo, New York : Cavendish Square Publishing, 2025. | Series: Inside guide: STEM superwomen | Includes bibliographical references and index.
Identifiers: LCCN 2024052427 | ISBN 9781502674562 (library binding) | ISBN 9781502674555 (paperback) | ISBN 9781502674579 (ebook)
Subjects: LCSH: Lovelace, Ada King, Countess of, 1815-1852–Juvenile literature. | Women mathematicians–Great Britain–Biography–Juvenile literature. | Women computer programmers–Great Britain–Biography–Juvenile literature. | Mathematicians–Great Britain–Biography–Juvenile literature. | Computer programmers–Great Britain–Biography–Juvenile literature.
Classification: LCC QA29.L72 J44 2025 | DDC 510.92 [B]–dc23/eng/20241213
LC record available at https://lccn.loc.gov/2024052427

Editor: Katie Kawa
Designer: Deanna Lepovich

The photographs in this book are used by permission and through the courtesy of: Cover Digital Image Library/Alamy Stock Photo; p. 4 Ada Lovelace portrait/Wikimedia Commons; p. 6 George Gordon Byron, 6th Baron Byron by Richard Westall (2)/Wikimedia Commons; p. 7 Annabella Byron (1792-1860)/Wikimedia Commons; p. 8 Powerloom weaving in 1835/Wikimedia Commons; p. 9 Miniature of Ada Byron/Wikimedia Commons; p. 10 Ada Byron aged seventeen (1832)/Wikimedia Commons; p. 12 Thomas Phillips - Mary Fairfax, Mrs William Somerville, 1780 - 1872. Writer on science - Google Art Project/Wikimedia Commons; p. 13 British (English) School - Charles Babbage (1792–1871) - 814168 - National Trust/Wikimedia Commons; p. 14 Massimo Parisi/Shutterstock.com; p. 15 Alan Spencer Photography/Alamy Stock Photo; p. 16 Tim Jenner/Shutterstock.com; p. 18 Anna Kepa/Shuterstock.com; p. 19 Stefano Mazzola/Shutterstock.com; p. 20 Diagram for the computation of Bernoulli numbers/Wikimedia Commons; p. 21 fizkes/ Shutterstock.com; p. 22 SeventyFour/Shutterstock.com; p. 24 chrisdorney/Shutterstock.com; p. 25 Popartic/Shutterstock.com; p. 26 adriaticfoto/Shutterstock.com; p. 27 Drazen Zigic/Shutterstock.com; p. 28 (top left) Charles Babbage 1860/Wikimedia Commons; p. 28 (top right) Alizada Studios/Shutterstock.com; p. 28 (bottom left) Ada Lovelace in 1852/Wikimedia Commons; p. 28 (bottom right) Bombardment of Fort Sumter/Wikimedia Commons; p. 29 (left) Ground Picture/Shutterstock.com; p. 29 (right) Gorodenkoff/ Shutterstock.com.

CPSIA compliance information: Batch #CSCSQ26: For further information contact Cavendish Square Publishing LLC at 1-877-980-4450.

Printed in the United States of America

CONTENTS

Ada Lovelace is often called the first computer programmer.

THE FIRST COMPUTER PROGRAMMER

Chapter One

Computer programs are a big part of our lives today. These sets of instructions allow a computer to solve problems in order to carry out tasks—from adding numbers to playing videos. The people who create these programs are called computer programmers.

We often think of computer programming as a modern career path. However, what many consider to be the first computer program was created in the 1800s! The programmer who created it was a woman named Ada Lovelace. She made a big difference in the world of STEM—science, technology, engineering, and math—at a time when women weren't given many opportunities to enter that world.

Famous Parents

Ada didn't come from a family of scientists. In fact, her father was the famous British poet Lord Byron. Her mother, Annabella Milbanke Byron, was a very smart woman who loved math. However, at the time, it was more acceptable for a woman to get married than it was for a woman to have a career, so that is what Annabella did.

Fast Fact

Computer programs are also known as software.

Ada's father, Lord Byron, was one of the most famous writers of his generation. His poetry is still widely read and studied today.

On December 10, 1815, Lord and Lady Byron welcomed their daughter, Augusta Ada Byron, into the world. This happy event could not help what was a troubled marriage, though, and the two separated not long after Ada was born.

The Importance of Logic

Ada's parents were very different people. Her mother was very logical, or driven by reason and facts. Her father was the opposite. He was often driven by his emotions, and it led him to do things that upset Annabella.

Annabella didn't want her daughter to be like that. She wanted Ada to be logical, so she provided her with tutors, or private teachers, to give her a good education.

Fast Fact

Throughout much of history, girls didn't go to school. During Ada's time, girls who had wealthy parents were often educated by tutors at home.

A Bright Child

Ada turned out to be like both of her parents. She had her mother's ability to

Ada's mother, who's shown here, had strong opinions and wasn't afraid to share them. For example, she spoke out against enslavement and fought for better conditions for people in prisons.

THE INDUSTRIAL REVOLUTION

Ada grew up during an exciting time for people with an interest in machines. The Industrial Revolution began in the 1700s and lasted through the 1800s. During this period, new machines allowed people to make goods more quickly and cheaply than ever before. This created new jobs in cities, so people moved from farms to cities in large numbers.

The Industrial Revolution began in Great Britian, which is where Ada lived, but it eventually spread to many different countries, including the United States. Factories became common in these countries, and in many cases, women were the ones who worked in these early factories.

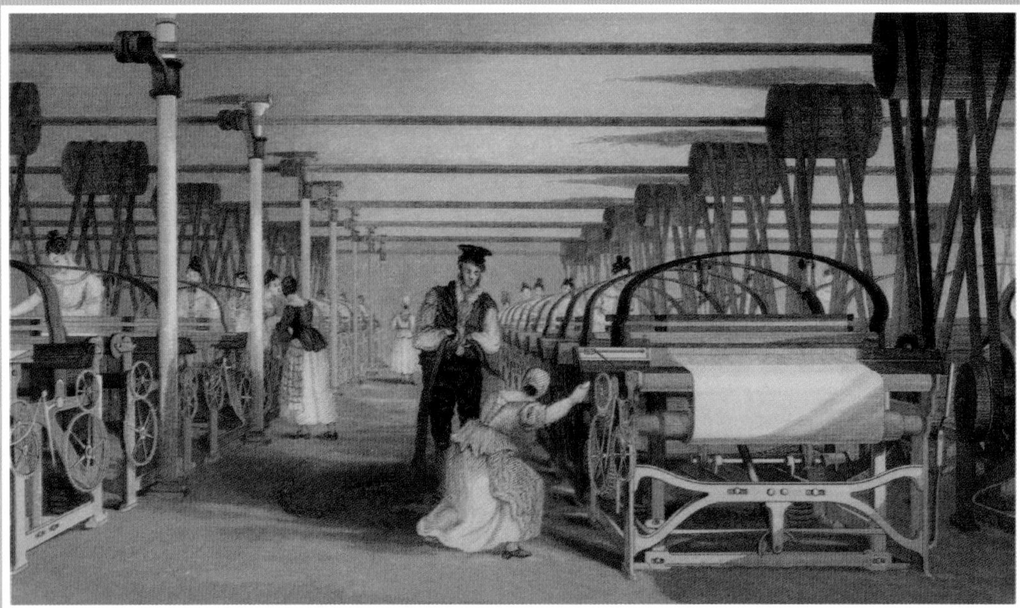

The Industrial Revolution first caught on in the textile industry, which dealt with the manufacturing of cloth. Shown here are women working in the textile industry in Great Britain in the 1800s.

understand math and science and her father's creativity. This became clear when she started imagining new inventions—including a flying machine!

When she was only 12 years old, Ada designed a steam-powered flying machine. She studied how birds were able to fly, and she used that knowledge to design her machine.

Growing Up

Ada was smart, creative, and interested in machines. She continued to learn from well-known teachers, and some of them believed she could have become a famous **mathematician**—if she had been born a boy.

In Ada's time, women weren't given the same respect or rights as men. Women were expected to get married and have babies—not become scientists or mathematicians. However, Ada soon met someone who showed her that women could do all those things and more.

Fast Fact

Ada was often sick. When she was 13 years old, one of her illnesses left her paralyzed, or unable to move. She had to stay in bed for a year!

Shown here is what Ada looked like as young girl.

When Ada was around 17 years old, she was introduced to society in London, England. This meant that she was taken to parties to let people know she was now old enough to get married.

IMPORTANT FRIENDSHIPS

Chapter Two

Ada didn't have a lot of friends while she was growing up. However, as she got older, she began making friends with others who enjoyed math and science as much as she did. Two of those friendships changed the course of her life.

The Queen of Science

Mary Somerville was one of Ada's tutors. Ada looked up to Mary because Mary was a respected scientist. She worked hard to educate herself even when others didn't support her. She studied many different branches of science, including geography, **physics**, and astronomy. In fact, she was called the "queen of science" after her death in 1872.

Mary and Ada became friends, and Mary would sometimes take Ada to lectures, or educational speeches, about science or math. She also introduced Ada to some of her friends who were scientists, including Charles Babbage.

Fast Fact

Mary Somerville's first husband did not support her love of learning. Her second husband, however, supported her and introduced her to other scientists.

Mary Somerville (*shown here*) and Ada Lovelace stayed friends for many years.

A Meeting of Mathematicians

In 1833, Mary introduced Ada to Charles Babbage. Charles was Mary's friend, and he was a respected mathematician. Charles and Ada met when she was 17 years old, which was when she was being introduced to London society. Although Charles was 42 at the time, he and Ada started a friendship.

MORE ABOUT MARY

Mary Somerville was born Mary Fairfax on December 26, 1780, in Scotland. When she was 10 years old, she went away to school for a year. After she came home, she continued to teach herself using books from her family's collection.

Mary eventually began writing papers about scientific topics. In 1831, she published a book about the solar system that made her famous. She went on to publish more science books, and she became one of the first women to be named an **honorary** member of Britain's Royal Astronomical Society—a group of some of the most respected scientists in the world.

Charles was interested in building a machine that could calculate, or compute, difficult math problems. In the 1820s, he began designing what he called the Difference Engine. He invited Ada to see a

Shown here is a portrait of Charles Babbage. His Difference Engine is believed to be the first example of a mechanical calculator—one that works using machine parts.

model of this machine, and she became very interested in how it worked.

The Difference Engine

Charles wanted to create a machine that could calculate mathematical tables. These were charts of answers to **complex** math problems that people could use to quickly get the answers they needed. However, because people were the ones figuring out these answers, the tables sometimes contained mistakes. This could be a big problem if someone was using a table to help them navigate at sea!

Charles believed the Difference Engine could

Fast Fact

Charles Babbage was one of the founders of the Royal Astronomical Society.

Babbage never finished building the Difference Engine because he ran out of money and moved on to his next idea. However, his designs were used to build the full machine after his death, and it is on display at London's Science Museum.

Fast Fact

The Difference Engine was designed to work in a way that is similar to today's calculators, but it would be able to do even more difficult math problems. It was also designed to store data like today's computers.

be used to **automatically** calculate these tables without mistakes.

Ada wasn't the only one who was interested in Charles's machine. The British government gave him money to work on it.

The Next Chapter

Charles was not the only man Ada met after being introduced to London society. In 1835, she married William King. When William was named Earl of Lovelace, Ada became the Countess of Lovelace, which is why she's commonly known as Ada Lovelace instead of Ada King.

Ada was a wife, and she soon became a mother. However, that didn't mean she gave up her interest in machines. In fact, a new chapter in her story as a computer scientist was about to begin.

Ada, William, and their three children lived in a part of England called East Horsley, shown here.

When Charles Babbage began sharing his plans for a new machine called the Analytical Engine (*shown here*), Ada used her love of machines and her creative imagination to think of many different ways it could be used.

CREATING A COMPUTER PROGRAM

Chapter Three

In some ways, Ada was following the path women of her time were expected to follow. In others, though, she was creating a new path for herself. That path led to Charles's latest machine and a world of possibilities for it that even he couldn't imagine.

The Analytical Engine

While Charles was working on the Difference Engine, he got an idea to create an even more advanced machine that could perform a wider variety of **calculations**. Instead of only calculating mathematical tables, this new machine could do all kinds of computing. Charles called it the Analytical Engine.

> **Fast Fact**
>
> The Analytical Engine is believed to be the first design for an all-purpose mechanical computer. This is why Charles Babbage is often called the "father of the computer."

In many ways, the Analytical Engine was designed to work like today's computers. Its mill was like a modern central processing unit (CPU) that can perform required tasks, and the store worked like memory does today. The Analytical Engine

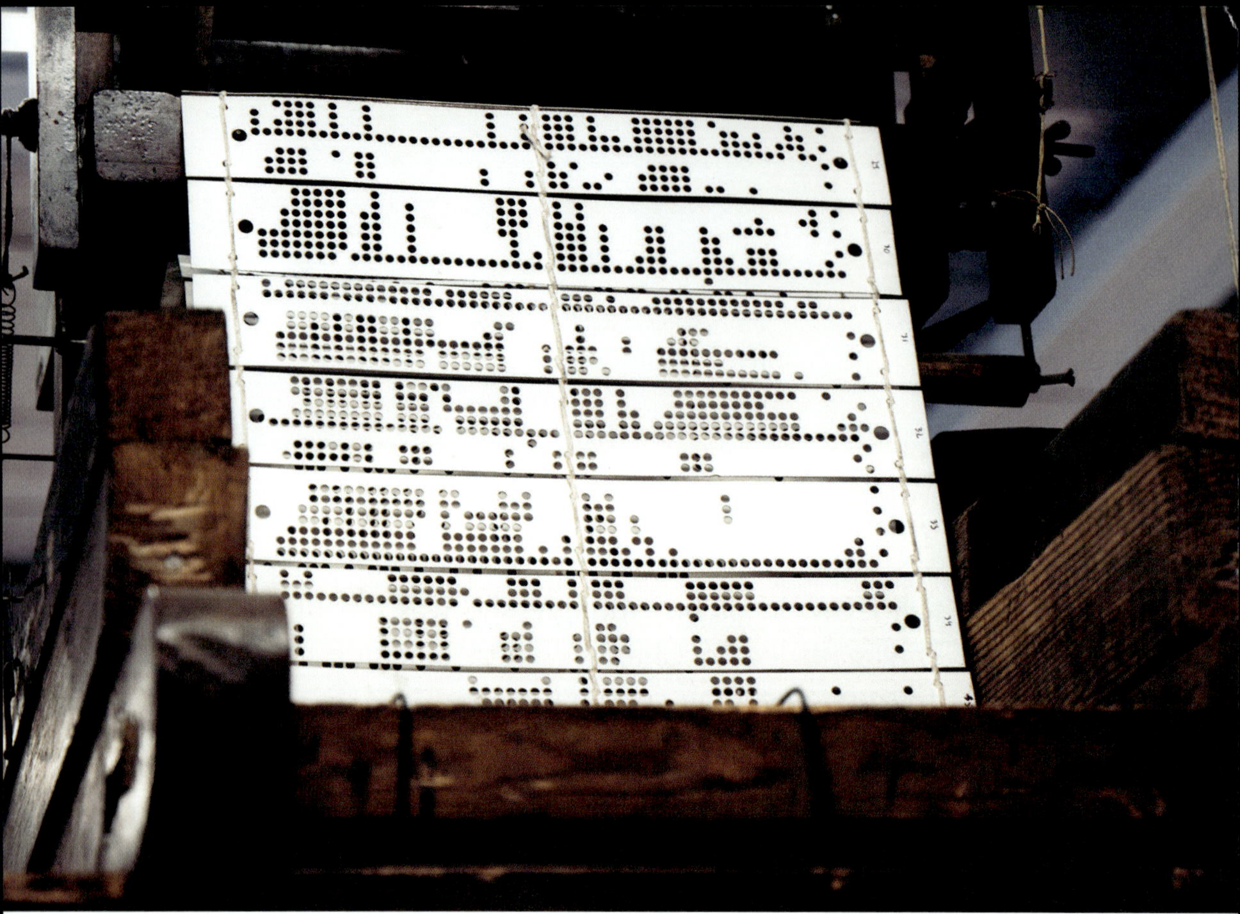

The punched cards that would tell the Analytical Engine what tasks to complete were based on those used in Joseph-Marie Jacquard's loom (*shown here*), which was used to weave cloth. The holes punched in the cards could be read by a machine as instructions.

would be given tasks to complete using punched cards, and it would print out the results.

Taking Notes

As Charles's friend, Ada learned a lot about the Analytical Engine. She had a strong understanding of how it worked. That understanding was put to good use when Ada was asked to translate a paper about the Analytical Engine from French into English.

ADA AND AI

One of the biggest issues in the world of computers today is the rise of artificial intelligence (AI). This technology enables computers to seemingly think and learn like humans.

In Ada's time, it seemed impossible for machines to learn or do anything without a person giving them direct instructions. She wrote, "The Analytical Engine has no pretensions [claims] to originate anything. It can do whatever we know how to order it to perform." The famous computer scientist Alan Turing called this "Lady Lovelace's Objection." He saw it as her objecting to, or disagreeing with, the idea that machines could ever think like a human.

Ada imagined many possible uses for computers, but even she couldn't imagine today's world of AI—a world where robots can bring people their food at restaurants!

Ada added her own notes about the Analytical Engine to her translation. Her notes were longer than the original paper! Ada's translation and her notes were published in 1843.

Fast Fact

The paper Ada translated was based on a speech Charles gave in Italy about the Analytical Engine.

Ada's Algorithm

One of the notes Ada left in her translation was an algorithm, or a set number of steps used to solve a math problem. She wrote that this algorithm could be used by the Analytical Engine to calculate Bernoulli numbers. These were a group of numbers often used in math that were difficult to calculate.

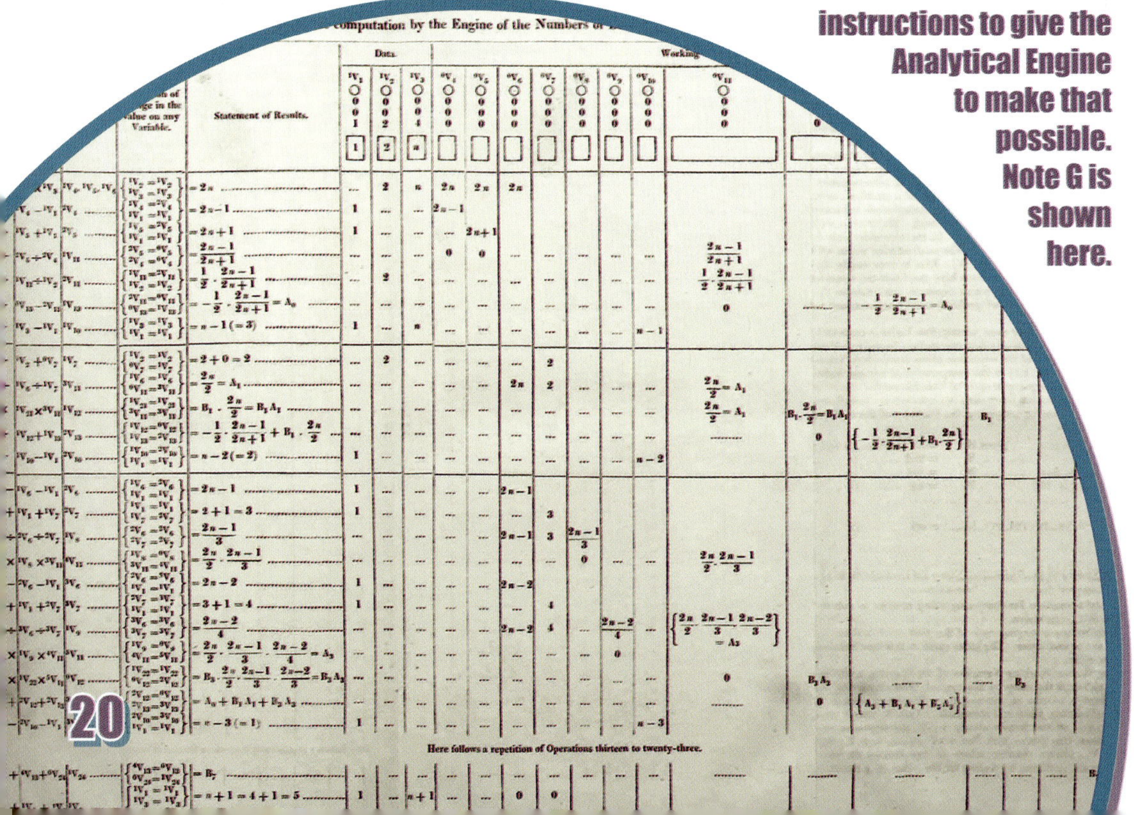

Ada gave her notes letters, and Note G described the algorithm for calculating Bernoulli numbers and the instructions to give the Analytical Engine to make that possible. Note G is shown here.

Ada's algorithm is believed to be the first ever created to be used by a computer. As such, it has been called the first computer program, making Ada the first computer programmer.

Fast Fact

Charles wrote his own algorithms for the Analytical Engine, but they were never printed like Ada's was. This is why Ada is often called the first computer programmer instead of Charles.

More Than Math

Ada had been combining her imagination and her love of machines since she was a girl designing flying machines. That mix of creativity and mechanical knowledge led her to think outside the box and imagine new possibilities for the Analytical Engine.

In her notes, Ada wrote that the numbers the Analytical Engine worked with could be used as symbols to stand for different things, such as music notes. This could allow the Analytical Engine to create music. Even Charles hadn't thought of such creative ways to use his invention!

Ada saw the potential for computers to create music, words, and even pictures.

Many women have followed in the footsteps of Ada Lovelace to become computer programmers.

Chapter Four

AHEAD OF HER TIME

Ada Lovelace played a huge part in the history of computers, but her work went unrecognized for more than 100 years. Then, as more people became interested in computers, her notes found a new audience who could fully appreciate them. In many ways, Ada was ahead of her time, and it took a while for the world to fully understand just how brilliant her ideas were.

A Life Cut Short

Ada lived a life full of big ideas, fascinating friendships, and new things to learn. However, that life was cut short on November 27, 1852, when Ada died after suffering from **cancer** for months. She was only 36 years old when she died.

Fast Fact

Charles Babbage never finished building the Analytical Engine. Like Ada, his work was forgotten for a century, until a new generation of computer scientists discovered it.

Ada didn't suffer alone, however. Her friends and family, including her mother, came to visit her. In fact, the famous writer Charles Dickens, who was one of Ada's friends, came

Fast Fact

Ada wanted to be buried next to the father she never knew, so she was buried next to Lord Byron at the Church of St. Mary Magdalene in the British town of Hucknall.

to her home to read to her when she was dying.

Rediscovering Ada

After Ada's death, her contributions to computer science were forgotten for many years. However, this began to change in the 1940s, when Alan Turing first mentioned her in a discussion of computer intelligence. Then, in 1953, B. V. Bowden shared Ada's notes on the Analytical Engine in a book he wrote about computers. After that, people began to take more of an interest in Ada and her work with computers. Computer programming was growing as a field of study, and more people began to recognize Ada's importance in the history of that field.

Great Britain takes pride in being the home of Ada Lovelace. This historical marker shows where Ada once lived in London.

Fast Fact

A programming language is the language used to give instructions to a computer. In the 1980s, the U.S. Department of Defense named its new programming language Ada, after Ada Lovelace.

Ada Lovelace Day

Ada is often seen as a role model for women and girls with an interest in STEM. That means she's someone they want to be like. Ada's influence on women and girls in STEM is celebrated every year on the second Tuesday of October—Ada Lovelace Day.

Ada Lovelace Day began in 2009 as a way to honor women in STEM and to support girls with an interest in these fields. On Ada

Ada Lovelace Day is celebrated in countries around the world. Its goal is to highlight the work of women in STEM, which can sometimes go unrecognized—just like Ada's work was for many years.

THE NEXT GENERATION

Many groups are working hard to support the next generation of girls who want to follow in Ada's footsteps and become computer programmers. Girls Who Code is an organization that was founded to teach girls the basics of coding, which is the part of computer programming that deals with writing the instructions for the computer to follow. Black Girls Code was created to inspire Black girls to become more involved in the world of technology. These groups and others like them are trying to increase the number of women in STEM careers, especially careers dealing with computers.

Today, 24 percent of all computer scientists are women. This number has gone down in the last 10 years. Many groups are working to get more girls interested in computers to help that number go back up in the future.

Lovelace Day, women working in STEM fields give talks about their work, and other events are held to make people more aware of the work women are doing in STEM.

Ada's Lasting Legacy

It can sometimes seem like the world of computer science is a man's world. However, the story of Ada Lovelace shows us that was never true. Women have always been involved in computer science—and other areas of STEM.

Ada Lovelace was more than just the world's first computer programmer. She was a woman who saw the potential for computers to do more than just math, and she was a woman who was proudly interested in math and science at a time when that wasn't always supported. This has made her a role model for girls and women around the world who love learning about STEM.

Ada Lovelace lived almost 200 years ago, but her influence on computer science—and on women and girls in STEM—is still felt today.

TIMELINE

In Ada's Life

1815
Augusta Ada Byron is born on December 10.

1833
Ada meets Charles Babbage.

1835
Ada marries William King.

1843
Ada's notes on Charles Babbage's Analytical Engine, which include the first documented computer program, are published.

1852
Ada dies on November 27.

World Events

1812–1815
The War of 1812 is fought between the United States and Great Britain.

1833
The Slavery Abolition Act is passed in Great Britain, which leads to the end of enslavement there.

1848
The Seneca Falls Convention is held in New York to call attention to the fight for women's rights.

1850
California becomes the 31st U.S. state.

1861–1865
The American Civil War is fought.

THINK ABOUT IT!

1. Use your knowledge of STEM and your imagination to be like Ada Lovelace and design a new machine. What would it do? How would it work?
2. What challenges do you think kept women from being more involved in STEM fields during Ada's time? Do those challenges still exist today?
3. Why do you think so few women today get jobs in computer science?
4. How would you celebrate Ada Lovelace Day?

GLOSSARY

automatically: Done through the use of machine parts that allow something to work without the help of a person.

calculation: The process of finding a solution to a problem using math.

cancer: A sometimes deadly sickness in which cells grow in a way they should not, often forming tumors, or growths, that harm the body.

complex: Not easy to understand or explain; having many parts.

data: Information that can be stored or sent through computers.

honorary: Given or done as a sign of honor.

mathematician: A person who is an expert in math.

physics: The branch of science that deals with matter and energy, as well as how they interact.

potential: Possibility.

FIND OUT MORE

Books

Lucas, Eileen. *Charles Babbage and Ada Lovelace: The Pen Pals Who Imagined the First Computer*. Buffalo, NY: Rosen Publishing, 2021.

Pierce, Nick. *Ada Lovelace*. New York, NY: Franklin Watts, 2020.

Sahai, Supriya. *Advancing Technology: Women Who Led the Way*. New York, NY: Children's Press, 2022.

Websites

Ada Lovelace Day
findingada.com/
Visit this website to find out more about Ada Lovelace and the day we celebrate every year in her honor.

Girls Who Code
girlswhocode.com/
The official website of Girls Who Code offers information on how this organization helps girls learn more about computers.

***National Geographic Kids*: Ada Lovelace**
kids.nationalgeographic.com/history/article/ada-lovelace
National Geographic Kids presents a short biography of Ada Lovelace.

Publisher's note to educators and parents: Our editors have carefully reviewed these websites to ensure that they are suitable for students. Many websites change frequently, however, and we cannot guarantee that a site's future contents will continue to meet our high standards of quality and educational value. Be advised that students should be closely supervised whenever they access the internet.

INDEX